CONTENTS

WHAT IS WIND POWER?

Wind power is a source of energy. It is used to run machinery, such as water pumps and electricity generators. Wind power causes less pollution than energy from fossil fuels or nuclear power. So it is less harmful to the environment.

Wind power turns turbines and pushes sailing boats. Most wind turbines are used for pumping water. But more and more are being used to make electricity.

Electricity made by wind power is more expensive than electricity made from fossil fuels. But it is getting cheaper every year.

◀ Wind turbines in California, in the USA. Electricity is carried away by the cables in the background.

FACTFILE

Wind turbines turn energy from the wind into energy that can do work. The wind turns the blades, which spin a horizontal shaft. The spinning shaft turns metal wires between magnets. This makes electricity. The power of a wind turbine is measured in kilowatts. One kilowatt is 1,000 watts.

▲ This simple windmill in Crete is used to pump water.

THE WIND

Winds are caused by the Sun heating the Earth unevenly. Air above hot land rises. It is replaced by air from cooler areas. These movements of air are the winds.

Winds are affected by the turning of the Earth. They are also affected by differences between the land and the sea. Land heats up and cools down faster than the sea, because moving water carries heat away.

FACTFILE

Clouds are a mass of tiny water droplets that float in the air. They form when warm, wet air rises up in the atmosphere, then cools and condenses (changes from a gas to a liquid).

Wind speeds

Winds travel at different speeds above the ground. Near the ground, winds are slowed by contact with the Earth. Between 10 and 15 kilometres above the Earth they form strong jet streams, which blow at 140 kilometres an hour. Some jet streams reach speeds of 450 kilometres an hour.

When warm air rises, cooler air takes its place underneath. ▼

Winds often blow along coasts. During the day, sea breezes blow from the sea to the land. Cooler sea air replaces warm air above land. At night, the breezes blow the other way.

Windsocks are used at airfields. They are tubes of material that are blown by the wind, and show pilots the direction and strength of the wind. ▶

The effect of the wind

Winds are extremely powerful. They create waves by blowing the sea's surface. The waves can be small ripples, or they can be great walls of water over 30 metres high.

Winds make sand dunes by piling up sand. They can also make dangerous sand storms, rain storms, tornadoes and hurricanes.

Air pollution in one country can be blown hundreds of kilometres by the wind. The wind also helps spread the Sun's heat more evenly through the Earth's atmosphere.

FACTFILE

The Moon has no wind to help spread the Sun's heat. So in the day, the surface becomes hotter than boiling water. At night, the temperature drops to 173 degrees Celsius below zero.

The wind has whipped up the sea around this ship, near the coast of Germany. ▼

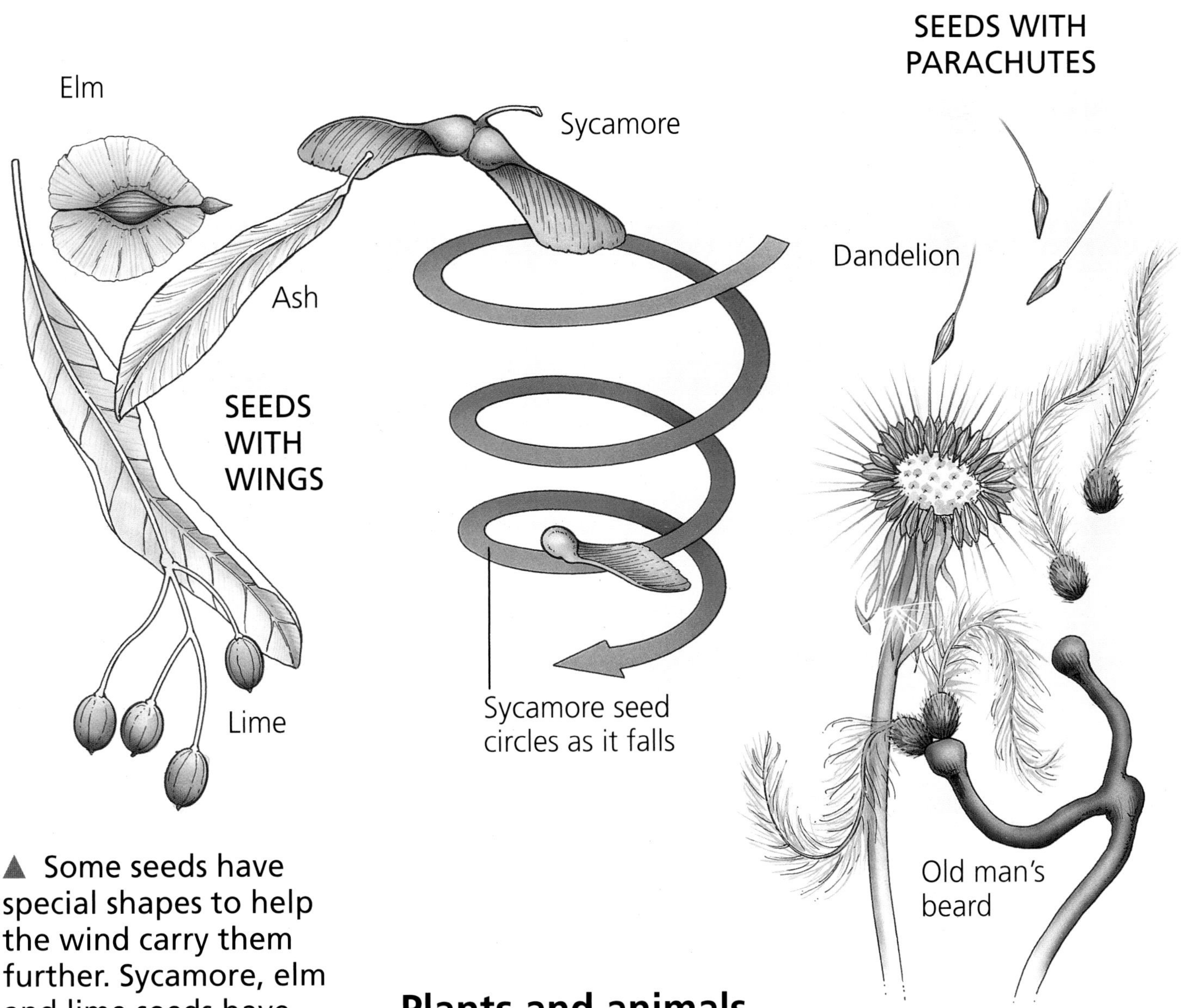

SEEDS WITH PARACHUTES

Elm

Sycamore

Dandelion

Ash

SEEDS WITH WINGS

Lime

Sycamore seed circles as it falls

Old man's beard

▲ Some seeds have special shapes to help the wind carry them further. Sycamore, elm and lime seeds have wings. Dandelion seeds have parachutes of feathers.

Plants and animals

Some plants use the wind to carry their seeds away. This helps the seeds find new food and light, away from their parent plants.

Some animals use the wind to help them find food. They can also smell predators many kilometres away, which gives them time to escape. The wind can also help them find a mate.

Where are the winds?

Wind changes from day to day, depending on the weather and the seasons. But there are patterns of wind direction and wind speed over the whole of the Earth.

Belts of wind blow towards the Equator, from the north and from the south. These are called trade winds. In the past, sailors relied on the trade winds to sail between the continents, trading goods.

FACTFILE

The main belts of wind in the world are the trade winds, the westerlies and the polar easterlies. The westerlies blow from the tropics towards the poles. The polar easterlies blow between the poles and the tropics.

◀ A kite festival in Dieppe, on the northern coast of France.

Wind turbines need wind speeds of at least 15 to 20 kilometres an hour. The best places to find these winds are on mountain passes, ridges, coasts or the shores of big lakes.

In the nineteenth century, ships called tea clippers used the trade winds to travel from China to Europe. ▼

The Beaufort scale

In 1805, a British admiral called Sir Francis Beaufort invented a scale from 0 to 12 for measuring wind speeds. In the Beaufort scale, winds are described by their effect on objects around them. For example, a light breeze measuring 2 on the Beaufort scale makes leaves rustle.

Tornadoes and hurricanes

Tornadoes and hurricanes are violent storms. Hurricanes form over warm seas. Spinning columns of wind can blow at over 200 kilometres an hour. When they hit a coast, hurricanes can cause massive destruction.

Tornadoes, or 'twisters', form on land under a thundercloud. Winds can reach over 500 kilometres an hour, picking up cars and buildings.

FACTFILE

Hurricanes, typhoons and tropical cyclones are all the same type of storm. They just happen in different places in the world.

This photo shows the spinning clouds of a hurricane. ▼

▲ Damage caused by Hurricane Marilyn, in the Virgin Islands in 1995.

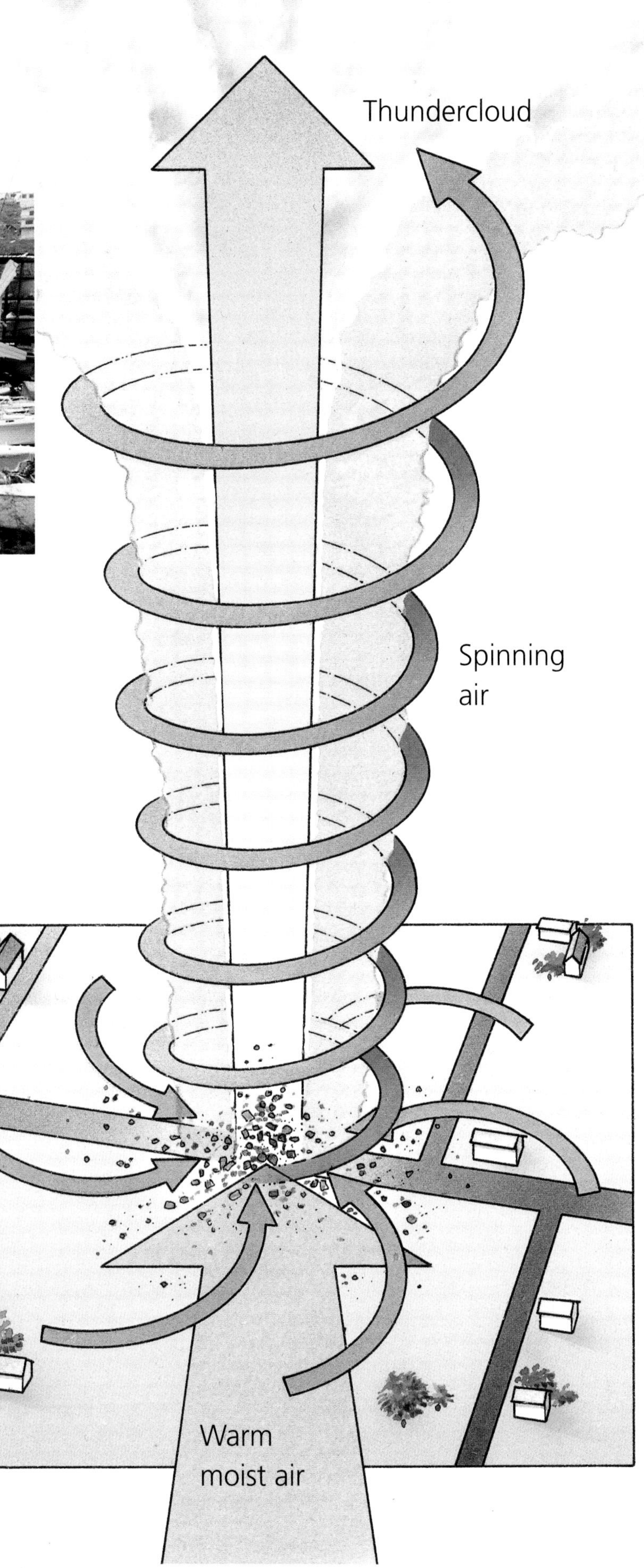

A diagram of a tornado. Warm, moist air rushes up to a thundercloud above. It spins and sucks in objects around it. ▶

HISTORY OF WIND POWER

FACTFILE

In 1588, strong winds wrecked over 25 ships from the Spanish Armada off the coast of Ireland. The ships had been beaten by the English in the North Sea and were trying to get home to Spain. The shipwrecks show what Spanish warships were like at that time.

Wind power was first used for sailing boats. Over 7,000 years ago, boats sailed down the River Nile in Egypt.

Later, sailing ships carried explorers around the world. They helped armies conquer new lands and empires grow. Ships also helped people trade goods between different countries.

Sails

The first sails were square. Ships had to travel with the wind right behind them. Then Arab sailors invented the triangular sail. It allowed ships to travel towards the wind, using differences in air pressure.

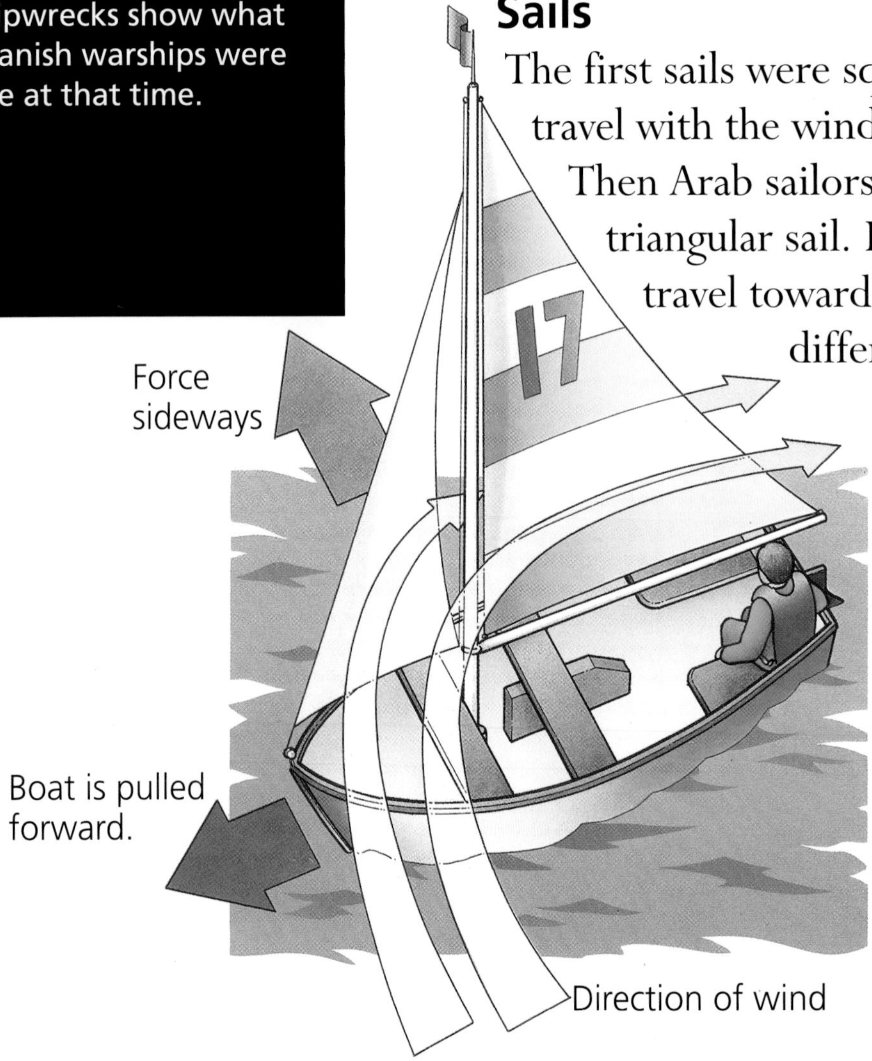

◀ Wind speeds up when it blows across a triangular sail. This makes the air pressure drop in front of the sail. That pulls the boat forward.

Boats called feluccas, on the River Nile in Egypt. Feluccas are still used today in the Middle East.

◀ The triangular sail was first used in the twelfth century, on an Arab boat called a dhow.

Wind machines

The first machines to use the wind were windmills. They were probably invented in the seventh century in Persia, which is now Iran. Later they were used in the Middle East, India and China.

Early windmills were used to grind grain. They had up to twelve sails hanging from a post. The moving sails of the windmill turned heavy millstones. The millstones ground, or squashed, the grain between them to make flour.

Inside a windmill, a wheel with teeth on it turned a smaller toothed wheel. The smaller wheel was linked to one of the millstones. ▼

A old type of windmill in Greece. The sails are triangles of cloth. ▶

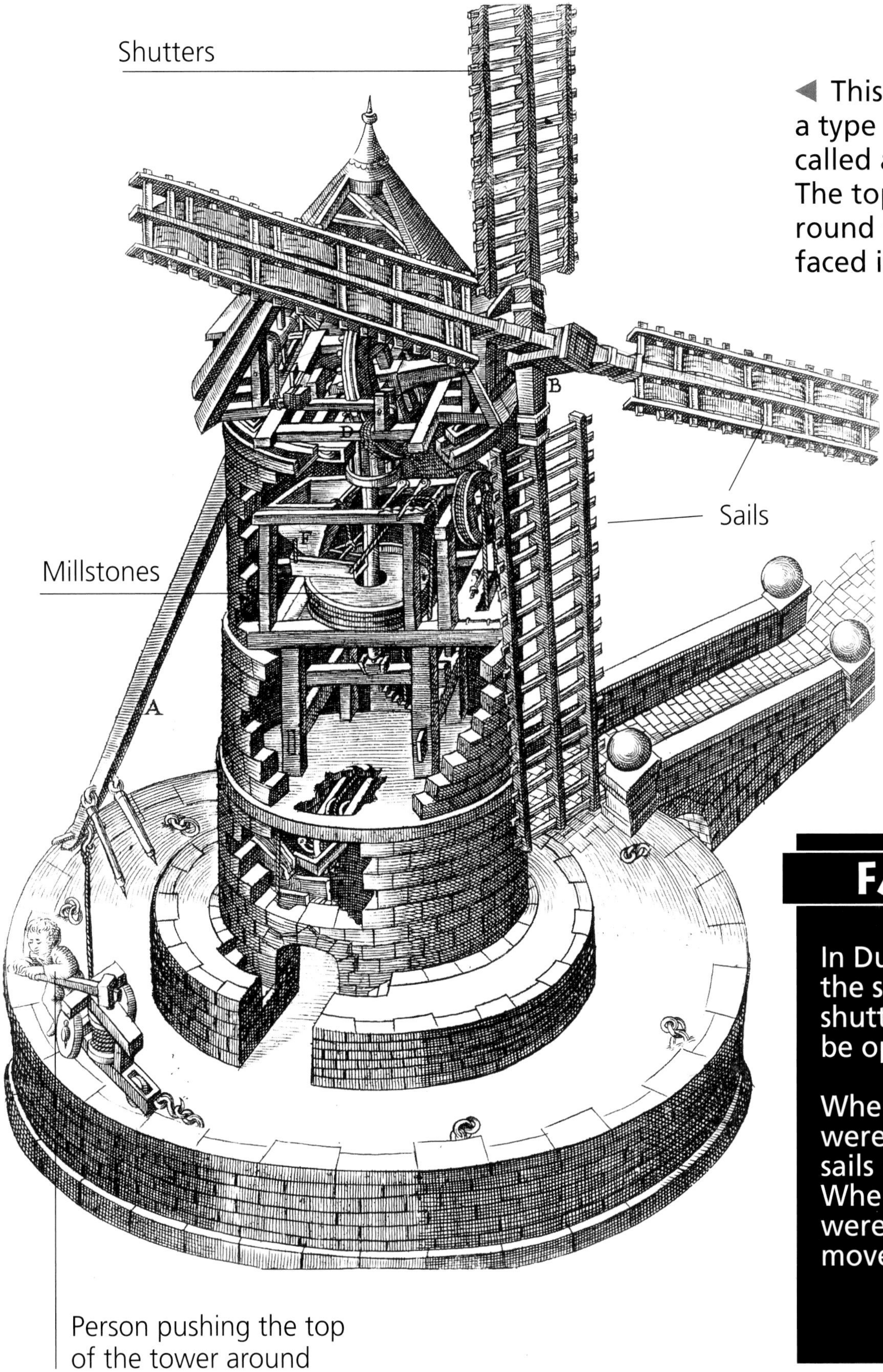

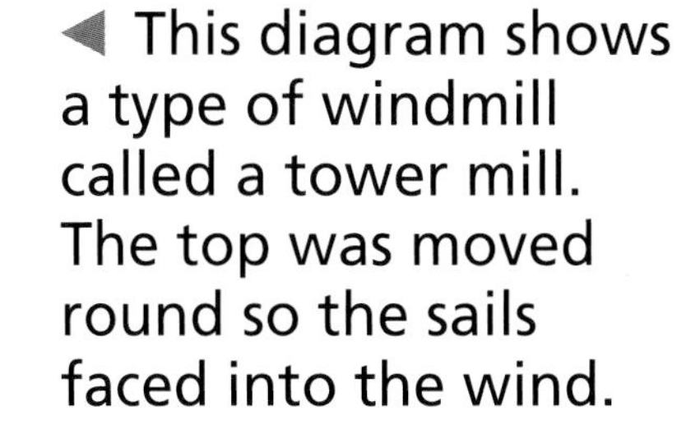

◀ This diagram shows a type of windmill called a tower mill. The top was moved round so the sails faced into the wind.

FACTFILE

In Dutch windmills, the sails had shutters that could be open or closed.

When the shutters were closed, the sails moved faster. When the shutters were open, the sails moved more slowly.

Windmills in Europe

In Europe, the first windmills, called post mills, were built in the twelfth century. To turn the sails to face the wind, the whole building was turned around. Later, the tower mill was invented. Only the top of the tower mill turned around. The top was fixed to the sails.

Windmills in the USA

In the middle of the nineteenth century, American farms started to use a type of windmill called a western wheel. It had blades instead of sails. When the blades turned in the wind, they drove a water pump. Western wheels are still used all over the world today. There is a photo of one on page 39.

An old post mill, in Sweden. The whole building was turned around by hand so that the sails faced the wind. ▶

FACTFILE

Windmills were built all over Europe. In the 1100s, the Dutch were using them to drain marshes and lakes near the sea. In the 1200s, some French towns had as many as 120 windmills around them. In the 1900s there were more than 700 windmills along the River Zaan in the Netherlands.

In the 18th century, fantails were invented. Fantails are small wheels, which turn by themselves to make the sails of the windmill face the wind.

WIND TURBINES

Wind turbines are modern windmills. Wind turbines that make electricity are also called aerogenerators.

There are two types of wind turbine. In the first type, the blades turn around a shaft that is parallel to the ground. These are called horizontal-axis turbines. Dutch windmills and American western wheels are like this.

In the second type of wind turbine, the blades turn around a shaft that stands upright. This is called a vertical-axis turbine.

Horizontal-axis turbines need hi-tech equipment to turn the blades into the wind. Vertical-axis wind turbines do not need to turn to face the wind. The blades work in any wind direction.

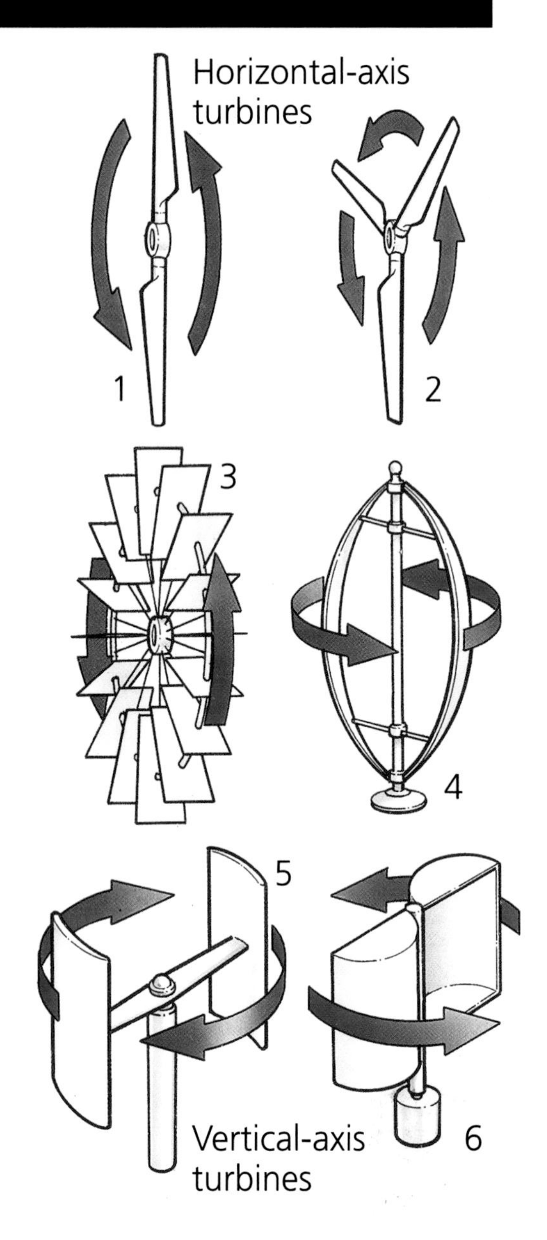

▲ The diagrams 1, 2 and 3 show blades on horizontal-axis turbines. Numbers 4, 5 and 6 show vertical-axis turbines.

Blades

Modern wind turbines have fewer blades than old windmills. They usually have only two or three blades.

FACTFILE

In the 1930s, a French inventor named Georges Darrieus designed a wind turbine that looked like a giant egg-beater. Today it is one of the most powerful turbines being used.

▲ A modern vertical-axis turbine, with two tall, thin blades.

This is a Darrieus wind turbine. The blades catch the wind from any direction.

Simple wind turbines

Many modern wind turbines use expensive, hi-tech equipment. But wind power can also be made using simple equipment made out of cheap materials. Poorer countries can use low-tech wind turbines to make power. This is called 'appropriate technology'.

A new type of wind turbine is tested in Britain. ▼

◀ This wind turbine has been made using a bicycle wheel.

Cheap materials

Wind turbines can be made with a bicycle wheel, with blades added to the spokes. Or they can be made from an oil drum.

If an oil drum is cut in half, the two halves become the blades of the wind turbine. This type of turbine is called a Savonius rotor.

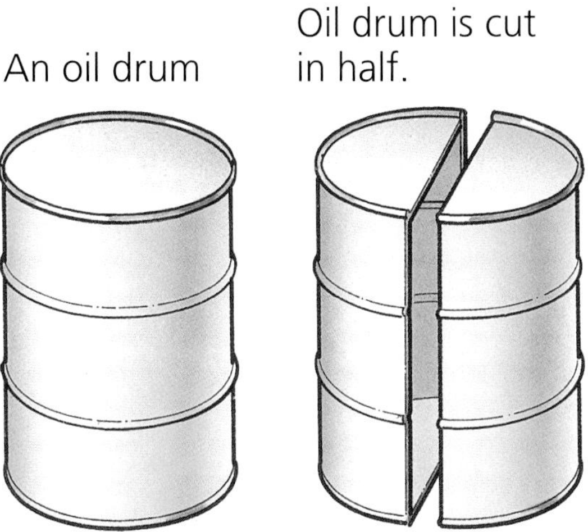

▲ To make a wind turbine out of two halves of an oil drum, the two halves are fixed to a steel pipe. The pipe slips over a pole. The wind fills the drums and pushes them round.

FACTFILE

Some advertising signs are shaped like the Savonius rotor on the right. As the sign spins, adverts on each half of the sign appear, one after the other.

Speed limit

As the wind speeds up, the blades of a wind turbine spin faster. After a while the blades would break off the turbine because of the powerful forces and vibrations made by the spinning. So turbines have special parts that limit their speed.

To spin slower, the blades of the turbine have to catch less wind. To do this, their angle to the wind can be changed. Or the whole turbine can be turned away from the wind.

Large wind turbines have hi-tech electronic controls, which constantly adjust the blades and the position of the turbine. Smaller turbines have weights fixed to the blades. As the blades spin faster, the weights change their angle so they catch less wind.

This turbine is being tested at the foot of the Rocky Mountains, in the USA.

As the blades of a wind turbine spin, strong forces pull the blades outwards. ▶

FACTFILE

The first big wind turbine was built in 1941, by an American engineer called Palmer Puttnam. It was built on top of a mountain in the USA. In 1945, one of its huge blades flew off in high winds.

Wind power and the environment

Wind turbines do not cause air pollution, so they are less harmful to the environment than fossil fuels. They also use a source of energy that will not run out, unlike coal, oil or gas.

Wind turbines do not take up a lot of space on the ground, so the land can also be used for growing crops or keeping animals.

Sheep graze under a row of spinning wind turbines. ▼

Noisy and ugly?

Some people think that wind farms are ugly. Others say that the swishing blades of the turbines make a lot of noise. Their gearboxes and generators can grind and clunk.

But new wind turbines are now being made with silent generators and quieter blades. So people who live near wind farms will not be disturbed.

FACTFILE

Electricity lines running across wind farms can kill many birds. The birds can find it hard to see the lines, and fly into them. When the farms are built, they disturb local wildlife. Sometimes the animals never return.

◀ These wind turbines in India were built by a Danish company. Local people are taught how to build and repair them.

USING WIND POWER

Electricity

To make electricity, a wind turbine has to have a generator. The blades of the turbine are joined to the generator, which is inside a cabin.

As parts of the generator spin, they move coils of wire past magnets. This makes electricity, which flows down the wire.

This part of a wind turbine joins one of the blades to the hub. ▼

Gearbox

In larger turbines used by electricity companies, the blades do not spin fast enough to drive the generator. So a gearbox is fitted between the blades and the generator to speed it up.

A gearbox has wheels with teeth fixed to a shaft. The shaft coming out of the gearbox spins faster than the blades. This shaft drives the generator.

FACTFILE

The biggest wind turbine in the world is in Hawaii. Its blades are 97.5 metres long.

▲ A crane lifts a generator cabin to the top of a turbine tower. It is on a wind farm in Germany.

The biggest wind turbines have generator cabins that are large enough to walk around inside. ▼

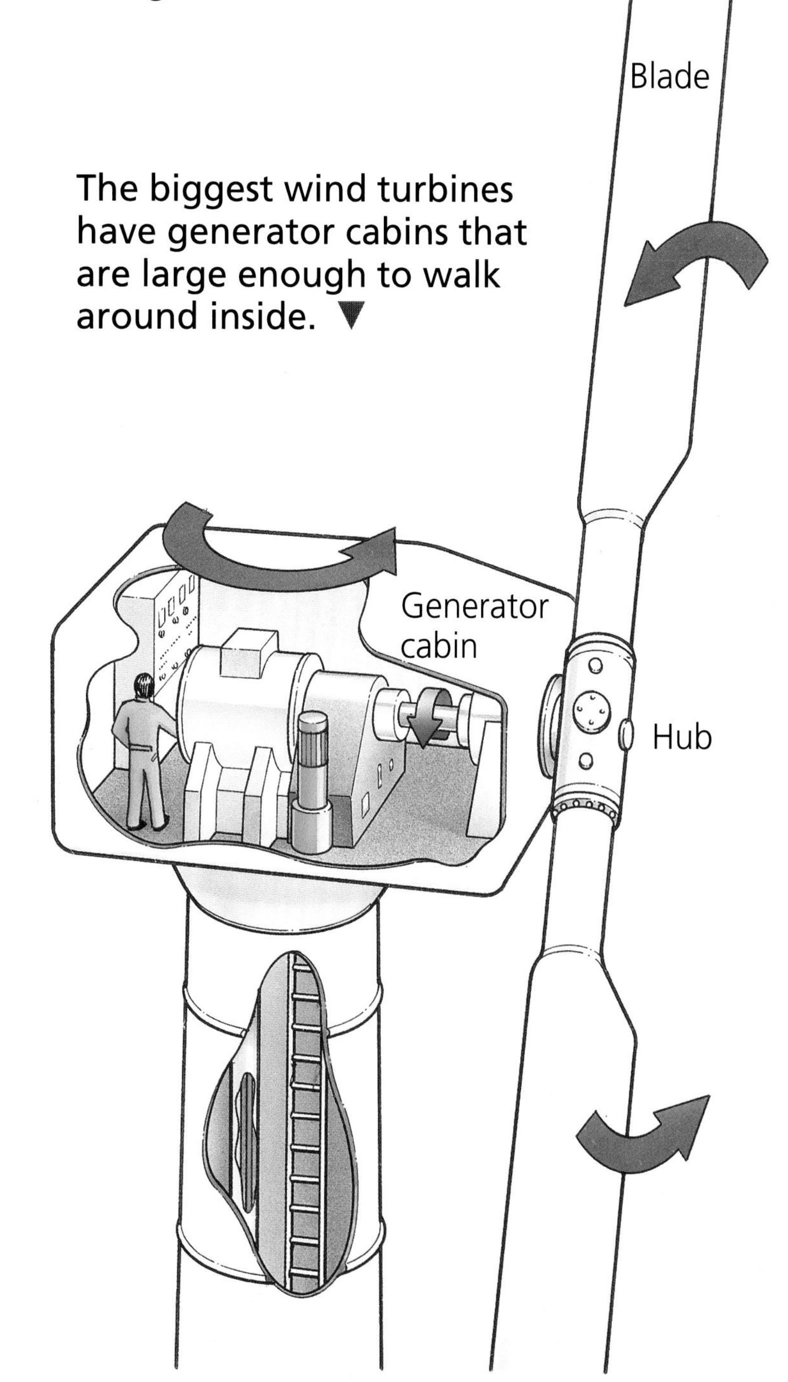

CASE STUDY · WIND POWER IN DENMARK

Denmark, the USA and Germany use the most wind turbines in the world. Denmark has over 3,700 wind turbines. They make over 3 per cent of the country's total electricity. This is about 1,000 million kilowatt-hours of electricity every year.

Most of Denmark's wind turbines are single turbines, or stand in a small group. They supply electricity to local towns and villages. Many are owned by local people rather than companies. Nearly 50,000 people in Denmark own or part-own a wind turbine.

Denmark also sells wind turbines abroad. It exports more turbines than any other country. Danish wind turbines are being used all over the world, from New Zealand and Mexico to China and India.

This power station in Denmark uses both coal and the wind to make electricity. ▶

These wind turbines use the sea breezes that blow along the coast.

An aerial photograph of a wind farm on the Danish coast. ▶

Home wind turbines

Small wind turbines can supply power to single homes. They are vital to people living in remote areas, where there is no other electricity supply.

In remote areas, the only other sources of power would be petrol-driven generators, or hi-tech solar panels. These are more expensive to run than wind turbines.

Home wind turbines can be made from kits. Or they can be made from scrap parts. Their blades are lightweight aluminium or fibreglass.

An alternator from an old car can be used as the electricity generator. Alternators make electricity to keep car batteries charged.

Small home wind turbines can make 750 watts of electricity. This is enough power for electric lights and small appliances like fridges.

FACTFILE

In 1991, an eco-village was built at Torup, in Denmark. Electricity is made by solar panels and a wind turbine. The wind turbine makes 450 kilowatts of electricity.

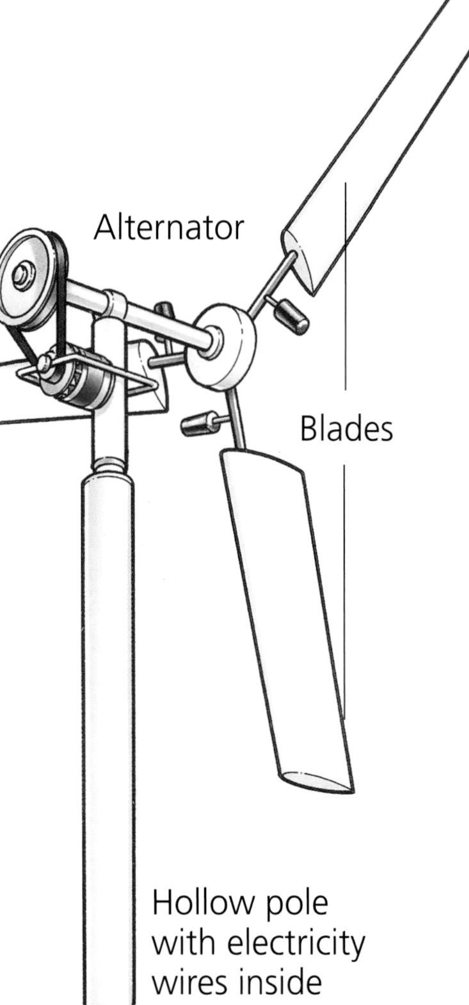

A small, home wind turbine sends electricity down to the ground along wires inside the pole. ▶

▲ Houses in the eco-village at Torup, in Denmark.

A wind turbine pumps water in an African village. ▶

Wind farms

Wind farms are large groups of wind turbines built by electricity companies. There can be hundreds or thousands of turbines on a single farm.

The turbines on a wind farm are massive. Their blades can be almost 100 metres across and their towers as high as 25-storey buildings.

A large wind farm can generate hundreds of megawatts of electricity. This is enough power for a whole town.

These wind turbines have open steel towers, which are quite cheap. They also blend in with the surroundings. ▼

Spaced out

The turbines on a wind farm have to be well spaced out. If they are too close, one can be in the wind shade of another.

The turbines are usually separated by about five times the distance across their rotors. They are carefully placed to get the most wind out of the smallest area.

FACTFILE

Building wind farms can disturb fragile habitats. The towers of the turbines need foundations up to 50 metres deep into the ground. When the deep holes are dug, dynamite is sometimes used to break up rocks. Rare species of plants can be destroyed.

This wind farm in Denmark was built on a slight hill. Winds often sweep up the sides of hills. ▼

◀ Rows of wind turbines in California. The turbine in the front of the photo has its generator housing open so it can be repaired.

The Altamount Pass in California, in the USA, has more wind turbines than anywhere else in the world. Over 6,000 turbines make enough power for about 100,000 homes.

▲ Two Darrieus 'egg-beater' turbines stand beside a horizontal-axis turbine.

Mountain passes are good places for wind farms. This is because the pass acts like a type of funnel. It squeezes the wind between the mountains, so it is faster and stronger.

The Altamount Pass is also near the coast. So when hot air rises above the land, cooler sea air is sucked in through the pass, making winds.

The US government plans to have more wind power in the USA. By 2020, it wants at least 5 per cent of the country's electricity to be made by wind power.

There are wind farms on islands like Hawaii. There are also wind farms on the windy coasts of Britain, Denmark and other Western European countries.

An old type of windmill on the Greek island of Crete. A triangular wooden vane fixed to the wheel swings the windmill into the wind.

Pumping water

Most wind turbines are used to pump water. The water is needed to irrigate farmland and for drinking.

Many parts of the world are too dry all year round to grow crops. Or there is a dry season, when there is no rain for a long time. Water is often just under the ground. Pumps are needed to bring it to the surface. Wind turbines make the power to drive the water pumps.

Before there were wind turbines, many people used kerosene to drive their water pumps. But these pumps often broke down, and did not last long. Wind turbines cost more to set up than kerosene-powered pumps. But once they are built, they are cheaper to run.

FACTFILE

Water is found underground in rocks. The level of water is called the water table. It rises when it rains, but it falls during dry times. When the water table falls, wind-powered pumps have to work harder to pump the water to the surface.

This fantail wind turbine in Argentina is also known as a western wheel. ▶

Wind-powered vehicles

Wind is still used to drive ships and other vehicles on the water. Most wind-powered craft are used for sport and leisure. There are yachts, sailing dinghies, windsurfers, ice-yachts, land yachts and kite-surfers.

Some passenger liners and oil-tankers are fitted with sails. They use wind power as well as engine power. When the wind is strong enough, they can turn off their engines.

Using wind power lowers the amount of fuel that is burned, so it makes journeys cheaper. It also reduces air pollution.

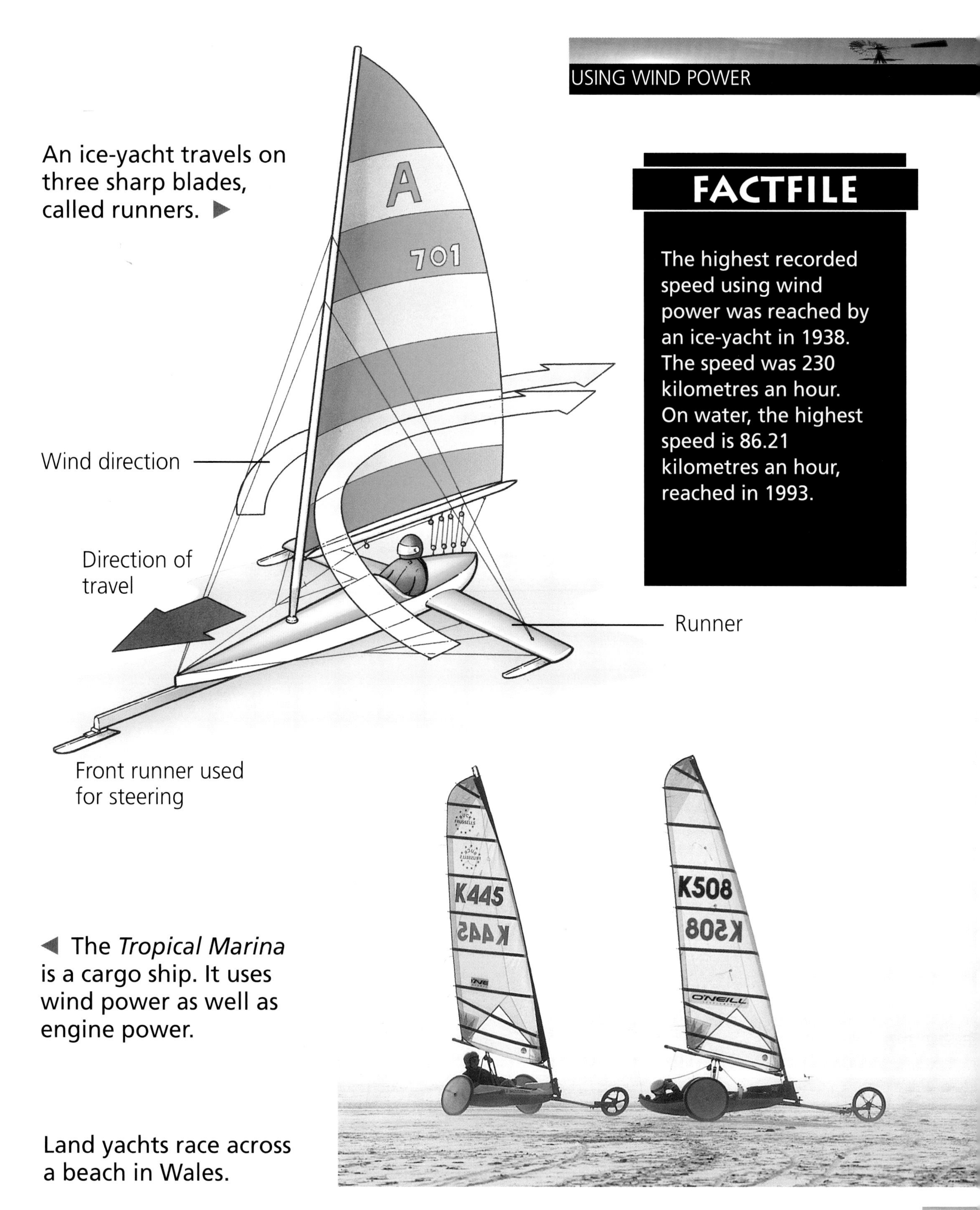

An ice-yacht travels on three sharp blades, called runners. ▶

FACTFILE

The highest recorded speed using wind power was reached by an ice-yacht in 1938. The speed was 230 kilometres an hour. On water, the highest speed is 86.21 kilometres an hour, reached in 1993.

◀ The *Tropical Marina* is a cargo ship. It uses wind power as well as engine power.

Land yachts race across a beach in Wales.

One of the biggest sailing ships in the world is called the *Club Med 1*. It is 187 metres long and weighs about 14,000 tonnes.

Club Med 1 is a luxury passenger liner, which made its first voyage in 1990. It has five masts and seven sails. The masts are made from aluminium. This is a very light metal. The masts are up to 50 metres high.

The sails are run by computers. They are continually adjusted so they make the most power from the wind.

The bottom of the ship's hull is only 5 metres below the surface of the water. This means it can enter shallow harbours, which many other passenger liners cannot do.

This photograph, taken from above, shows some of the eight decks of *Club Med 1*. ▼

◀ Passengers on the deck of *Club Med 1*.

This luxury sailing yacht is called *Wind Song*. It sails around the Mediterranean and Caribbean Seas. An interested dolphin is swimming closer to have a look.

FACTFILE

In the past, sailors had to climb up the masts to adjust the sails. But now, computers on board the *Club Med 1* run electric motors to adjust the sails.

THE FUTURE

Wind power is a free, renewable resource. It will never run out. It is also a clean source of electricity. Many countries are building more wind turbines to increase the amount of wind power they use.

Between 1998 and 1999, the world's wind power increased by 36 per cent because of new turbines.

This diagram shows how wind power could be used in the future, to make electricity for cars, ships and homes. ▼

FACTFILE

Offshore wind turbines are specially designed to keep out the sea water. This is because salt water corrodes metal. The towers of offshore wind turbines are airtight. In their generator cabins the air is at a higher pressure than normal, to stop the wet sea air seeping in.

Offshore wind turbines

Wind farms are now being built in the sea, where the winds are often stronger. In 1991, Denmark built 11 wind turbines in the sea near the Danish coast. They make 5 megawatts of electricity a year.

By 2010, Britain plans to make 10 per cent of its electricity from wind power and other renewable energy sources. As engineers find ways of making it cheaper, wind power will become more important in the future.

A wind farm at Ebeltoft, on the coast of Denmark. Many more farms like these will be built in the future.

GLOSSARY

Appliances Tools or small machines used in homes.

Atmosphere The mass of gases that surround the Earth.

Axis A straight line that an object turns around.

Cargo The load of goods carried by a ship or an aircraft.

Condenses Changes from a gas into a liquid by cooling.

Corrodes Wears away gradually.

Environment Everything in our surroundings.

Exports Sells and sends to countries abroad.

Fossil fuels Coal, oil and natural gas formed from the remains of plants and animals over millions of years.

Generator A machine that changes movement energy into electricity.

Habitats Natural homes of plants and animals.

Hi-tech Advanced, or new, technology.

Horizontal Parallel to the horizon, or side to side.

Hub The central part of a wheel.

Irrigate To supply with water.

Jet streams Currents of air travelling at very high speed 10 -15 kilometres above the ground.

Kerosene A cheap fuel, made from oil.

Kilowatt (kW) One thousand watts, a measure of electrical power.

Kilowatt-hours (kWh) A unit of energy that is equal to 1,000 watts of electrical power being used for one hour.

Liner A large ship.

Millstones A pair of rounded, flat stones used for grinding corn, wheat or other grain.

Predators An animal that hunts another, usually for food.

Renewable Something that can be replaced.

Rotation The movement of turning around a centre.

Scent Smell.

Shaft A bar that helps parts of a machine turn.

Turbines Angled blades fitted to a shaft that are made to turn by the force of water, steam or air.

Vertical Straight up and down.

FURTHER INFORMATION

Books to read

Action for the Environment: Energy Supplies by Chris Oxlade and Rufus Bellamy (Franklin Watts, 2004)

Alpha Science: Energy by Sally Morgan (Evans, 1997)

A Closer Look at the Greenhouse Effect by Alex Edmonds (Franklin Watts, 1999)

Cycles in Science: Energy by Peter D. Riley (Heinemann, 1997)

Essential Energy: Energy Alternatives by Robert Snedden (Heinemann, 2002)

Future Tech: Energy by Sally Morgan (Belitha, 1999)

Saving Our World: New Energy Sources by N. Hawkes (Franklin Watts, 2003)

Science Topics: Energy by Chris Oxlade (Heinemann, 1998)

Step-by-Step Science: Energy and Movement by Chris Oxlade (Franklin Watts, 2002)

Sustainable World: Energy by Rob Bowden (Hodder Wayland, 2003)

Power station produces several million watts.

Family house uses a few thousand watts.

Washing machine: 2,500 watts

Electric iron: 1,000 watts

Light bulb: 100 watts

ENERGY CONSUMPTION

The use of energy is measured in joules per second, or watts. Different machines use up different amounts of energy. The diagram on the right gives a few examples. ▶

INDEX